Experimenting WITH Air

Gordon R. Gore

Trifolium Books Inc.
Toronto, Canada

Trifolium Books Inc.
250 Merton Street, Suite 203
Toronto, Ontario, Canada M4S 1B1
Tel: 416-483-7211 Fax: 416-483-3533
e-mail: trifoliu@ican.net

We acknowledge the financial support of the Government of Canada through the Book Publishing Industry Development Program (BPIDP) for our publishing activities.

Canada

Canadian Cataloguing in Publication Data

Gore, Gordon R.
Experimenting with air: hands-on science activities

ISBN 1-55244-042-7

1. Air—Experiments—Juvenile literature. I. Title. II. Title: Air.
QC161.2.G67 2000 533'.6'078 C00-930556-4

Please note: In most cases, the terms and units in this book have been presented in standard S.I. style. In some instances, however, the author has deviated from S.I. in order to clarify ideas, reinforce new terms, and/or improve the flow of the text.

Project editor: Sara Goodchild
Design and layout: Heidy Lawrance Associates
Project coordinator: Sara Goodchild
Production coordinator: Heidy Lawrance Associates
Cover design: Heidy Lawrance Associates
Cartoons: Ehren Stillman; Computer line drawings: Moira Rockwell

Printed and bound in Canada
10 9 8 7 6 5 4 3 2 1

Trifolium's books may be purchased in bulk for educational, business, or promotional use. For information, please contact: Special Sales, Trifolium Books Inc., 250 Merton Street, Suite 203, Toronto, Ontario, M4S 1B1 Tel: 416-483-7211

Introduction

Experimenting with Air is a book of hands-on activities for young people. The activities may be done by pairs or small groups of students. Many of the activities require only readily-available materials.

The author has attempted to capture the hands-on spirit of the pan-Canadian science curriculum. This book is intended to be just one useful resource for the Air and Flight part of that curriculum. Teachers will no doubt use many resources, print and non-print, to help them achieve the many stated learning objectives of the pan-Canadian curriculum.

In the author's view, the main objective of an elementary science unit is to build on the natural interest students have in their physical world by providing them with as much concrete, truly hands-on experience as possible. Hands-on science makes the subject real and enjoyable.

The author hopes this book will be used to let students do some science, and learn about the properties of air through direct experience. If, when they finish using this book, students are eager to learn more about air and flight, or science in general, then my main objective for elementary science will have been achieved.

Acknowledgments

Thank you to my favorite cartoonist Ehren Stillman, of Abbotsford, B.C., for contributing the cartoons that appear in this book. Ehren's work has been published in many science and physics books by this author, in two chemistry texts by Dr. Jim Hebden, in *The Physics Teacher* (New York) and in *The Science Teacher* (Washington, D. C.). Most of the computer line drawings in this book and in the teacher's guide are by Moira Rockwell of Mission, B.C. Her fine work is always greatly appreciated. All photographs are by the author.

Safety

Introduction to Safety in the Science Lab

The activities in this book have been teacher and student-tested, and are safe when carried out as directed, with proper care. These activities should not be attempted without the permission and supervision of a teacher.

When special attention to safety is required, an icon like this ⚠ appears beside the activity.

In addition to the routine safety procedures used in your school, there are some general guidelines that are important to follow when working through any science activity.

Safety Guidelines for Students

When you begin

- Listen carefully to your teacher's instructions.
- Make sure your teacher knows if you have allergies or any other medical or physical conditions that might affect your ability to carry out activities safely.
- If you wear a hearing aid or contact lenses, tell your teacher before starting any activities.
- Start working on activities only after your teacher tells you to begin. Get your teacher's permission before changing any steps in an activity.
- Know the location and proper operation of blanket, fire extinguisher, fire alarm and any other emergency equipment.

In the Lab

- Wear a lab apron, safety goggles, and/or closed-toe shoes when your teacher directs you to do so.
- Make sure long hair or loose clothing is tied back.
- Never eat or drink while you are working.
- Work quietly and carefully with a partner, and pay attention to the actions of others around you as well as your own.
- Inform your teacher immediately of any damaged equipment or glassware, any accident, or any behavior that you consider dangerous.

Experiments using Electricity

- Your work area and your hands should be dry when in contact with any electrical equipment.
- Always unplug electrical equipment by pulling the plug, not the cord.
- Report any damaged equipment to your teacher.
- Make sure electrical cords are out of the way so that you or others will not trip over them.

Experiments using Heat

- Always use heatproof containers that are whole and undamaged.
- Never use a Bunsen burner or hot plate without the permission and supervision of your teacher.
- Direct the open end of a container that you are heating away from yourself and your classmates.
- Never allow containers to boil dry.
- If you burn yourself, inform your teacher and cover the burned area with cold water or ice immediately.

Experiments using Chemicals

- Handle substances carefully, and do not touch substances unless you are instructed to do so.
- If you are instructed to smell a substance, never smell it directly. Instead, hold the container slightly away from your face and waft fumes towards your nose with your hands.
- Always hold containers away from your face when pouring liquids.
- If any part of your body touches a substance, rinse the area immediately and continuously with water. If something gets into your eyes, do not touch them. Immediately rinse them with water for fifteen minutes, and inform your teacher.

Cleanup

- Wash your hands after completing any activity.
- Clean all equipment and put it away according to your teacher's instructions.
- Follow your teacher's instructions for cleaning up spills and disposing of materials.

Table of Contents

SECTION 1 • Is Air Real Matter?

We are surrounded by air. We cannot see it, we cannot smell it, and we cannot feel it unless there is a wind. How do we know that air is really 'matter'? What are some of the properties of air? How do we use these properties in everyday life? The activities that follow will help you answer these questions.

1.1 Try This Yourself! What's in the Bag?

What You Need

1 large, clear, garden-size garbage bag

What to Do

1. Fill a large, clear, garden-size plastic bag with air. Gather up the open end of the bag in one hand, so that you trap air inside the bag. See **Figure 1.**
2. Give the bag of air a quick jerk in one direction. Watch what happens to the air in the bag. Does the air 'go willingly' with the bag, or does it tend to 'stay where it was'?
3. Get the bag moving at a steady speed, then stop the bag quickly. Does the air inside the bag 'stop willingly,' or does it tend to keep on moving?

Figure 1

Questions

1. When you jerked the bag of air to one side, did the air inside the bag resist change in its motion? How could you tell?
2. When you moved the bag of air, then stopped the bag quickly, did the air inside the bag resist change in its motion? How could you tell?

EXERCISE CARE WHEN USING PLASTIC BAGS

Inertia

Anything made of matter will resist change in its motion. If an object is still, it will tend to stay still. If the object is moving, it will tend to keep moving at the same speed and in the same direction. This property of matter is called inertia.

Mass and Inertia

Anything with mass has inertia. In fact, when you measure mass, you measure inertia. All matter has mass, therefore all matter has inertia. When you measure your own mass in kilograms, you are also measuring how much inertia your body has.

Question

According to your observations on the bag of air, does air have mass? How could you tell?

1.2 Try This Yourself! Take the Plunge!

What You Need

1 kitchen plunger (previously unused)

What to Do

1. Hold a clean bathroom plunger in your hand, just to get an idea of how heavy it is.
2. Moisten the plunger (**Figure 2**) and push it down firmly against a *smooth* table top, so that most of the air is squeezed out from under the plunger.
3. Now, try to pull the plunger off the table, by pulling straight up. Why is it so difficult to do this?

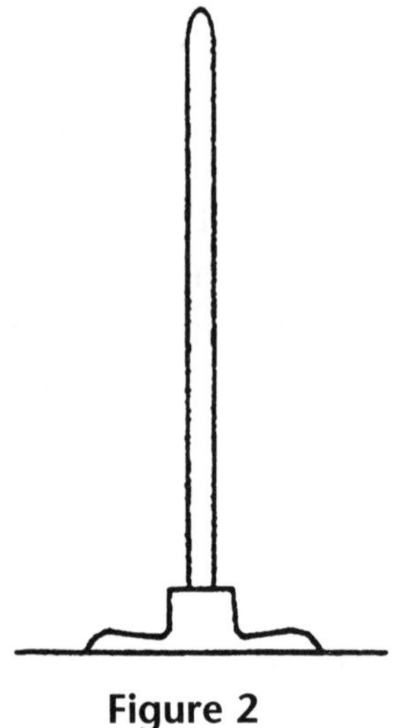

Figure 2

Mass and Weight

If something has mass, it will be pulled toward Earth by a force we call gravity. The amount of force the earth exerts on a mass is called its weight. If air surrounding Earth has mass, it must also have weight.

We are surrounded by an ocean of air, which reaches to the top of the atmosphere. When you hold a plunger in your hand, the air around the plunger pushes on its surface in all directions. When you push the plunger down on the table, you squeeze most of the air out from under the plunger. The space underneath the plunger is now almost a vacuum. However, there is still lots of air above the plunger! It is the weight of the column of air above the plunger that makes it so hard to lift the plunger off the table.

Questions

1. When you push the plunger down onto the table, does this change the weight of the plunger itself?
2. Why is the plunger so hard to lift off the table after you squeeze the air out from under it?
3. Workers installing windows in buildings may use rubber suction cups, with handles, to move large panes of glass around safely. What keeps the suction cups from falling off the glass windows?
4. Why does the plunger experiment suggest that air has mass?

Volume

If air is matter, then not only must it have mass, it must also occupy some space. The amount of space a material takes up is called its volume. You have just held a bag of air in your hands, so it seems that air occupies volume. The next activity should convince you even more that air has volume.

1.3 Try This Yourself! How Can You 'Pour' Air?

Figure 3 **(a)** **(b)**

What You Need

2 drinking glasses, or 2 250-mL beakers
1 large, unoccupied aquarium, or a deep sink, filled with water

What to Do

1. Fill a large bucket, empty aquarium, or sink with water. Push an 'empty' glass straight down into the water, as in **Figure 3(a)**.
2. Push a second glass down into the water, but fill it with water. Turn it upside down, too. See **Figure 3(b)**.
3. Now, see if you can pour air from the first glass up into the second glass.

Question

Does this activity suggest that air occupies volume? How could you tell?

1.4 Try This Yourself! Make a Tornado

What You Need

2 2-litre plastic pop bottles
1 Tornado Tube™ (bottle connector)

What to Do

1. Connect two used pop bottles using a Tornado Tube™ connector. One of the bottles is about ¾-full of water.
2. When you put the water-filled bottle on top, very little water will flow down into the empty bottle below. Try 'swirling' the bottles until you create an air passage so that air can pass from the bottom bottle up into the top bottle. Now water will flow easily down through the connector, and you have a miniature 'tornado' to watch! See **Figure 4.**

Figure 4

Questions

1. What material is in the bottom bottle, when the bottle full of water is placed on top?
2. Before you 'swirl' the bottles, why do you think it is difficult for the water in the top bottle to flow down into the bottom bottle?
3. How does this activity suggest that air is matter?
4. Why does the water flow smoothly down into the bottom bottle after you 'swirl' the bottles?

What Is Air?

We live at the bottom of a huge ocean of air. Earth is the only planet in our solar system that is surrounded by an atmosphere that supports life as we understand it. Air is really a mixture of several gases. The two most abundant gases in our atmosphere are nitrogen and oxygen. Oxygen is absolutely essential to life as we humans know it. There are small quantities of other gases in air.

What the Experts Say about Air

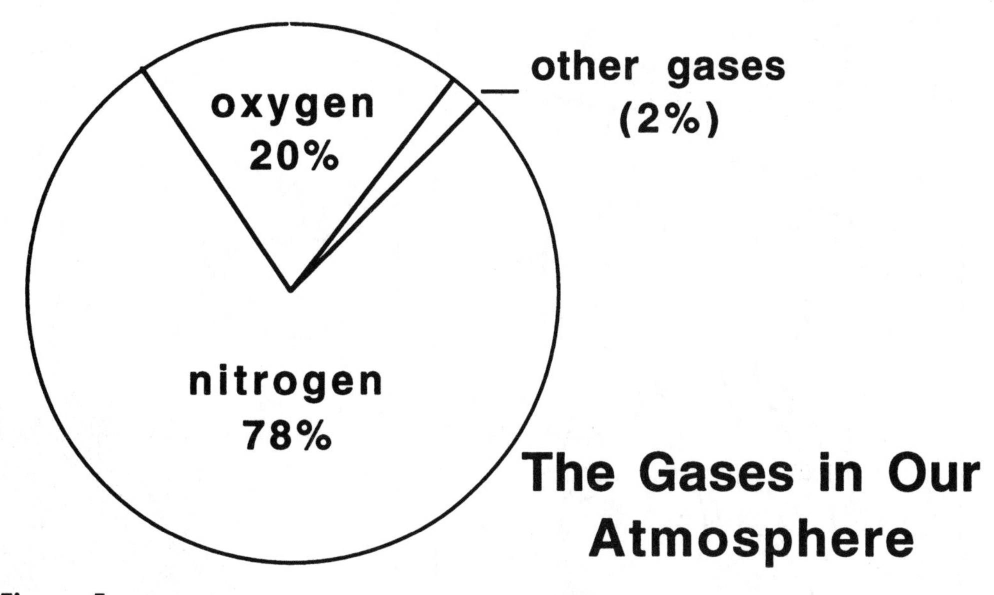

Figure 5

Scientists who analyze the contents of air in the lower atmosphere have found that about 20% of dry air is oxygen gas. Most of the air is nitrogen gas (78%). The remaining 2% of dry air is made up of argon, carbon dioxide, and traces of methane, hydrogen, and other gases (see Figure 5). Air also contains water vapour. The amount of water vapour varies from day to day and from location to location.

SECTION 2 • Air Pressure

Weather forecasters talk about air pressure when they predict the weather. High pressure usually means 'good' weather (clear skies), and low pressure can mean 'bad' weather (cloudy skies). What causes air pressure? First, we have to explain what **pressure** means.

Figure 6 (a) (b)

What is Pressure?

Imagine you are trying to lift a large pail of water, using one finger wrapped around the thin handle of the pail. See Figure 6(a). You can hold the pail for a short time, but soon it becomes very uncomfortable.

You put the pail down, then lift it using four fingers. See Figure 6(b). You can now hold the pail quite comfortably. The force of gravity on the pail of water is the same as before. Why is it easier to carry the pail with four fingers than it was with one finger? With four fingers, the force of gravity on the pail is spread over four times the area that it was with one finger. This means that the pressure on each of your four fingers will be about one quarter of what it was when all the force was exerted on one finger.

Pressure is the amount of **force** divided by the **area** over which the force acts.

$$\text{Pressure} = \frac{\text{Force}}{\text{Area}}$$

When skaters put all their weight on one skate, they exert a very high pressure on the ice. This is because the area over which the force of gravity acts is very small. If you divide the force of gravity by a small number, you obtain a large number for the pressure. The high pressure a skater exerts on the ice makes it melt, and the skate blade actually slides on a thin layer of water. Once the skater passes over a point in the ice, the ice refreezes.

Sometimes it is important to reduce pressure. All-terrain vehicles (ATVs) have very wide tires. This distributes the force of gravity on the ATV over a larger area. Because the area over which the weight is spread is larger, the pressure on the ground is smaller. With their wide tires, golf carts and ATVs exert very little pressure on the ground, and therefore do little damage to the land.

$$\frac{\text{Force}}{\text{Small Area}} = \text{High Pressure}$$

$$\frac{\text{Force}}{\text{Large Area}} = \text{Low Pressure}$$

Area = 1 m^2

Figure 7

Scientists measure pressure in a unit called the pascal (Pa). One pascal represents a very low pressure. The pressure exerted on us by our atmosphere is about 101,000 Pa.

Because the pascal is such a small unit of pressure, the weather forecaster measures atmospheric pressure in kilopascals (kPa). A kilopascal is equal to 1000 Pa, therefore atmospheric pressure is about 101 kPa.

Atmospheric pressure is caused by the weight of air above you. If you climb or fly to a higher altitude, atmospheric pressure will be less, because there will be less air pushing down on you from above. In Figure 7, atmospheric pressure will be less for the aircraft than it will be for the person standing on earth's surface.

Atmospheric pressure at sea level is equal to the weight of about 200 grade six students stacked up on an area of one square metre!

We are so accustomed to living in our atmosphere that we take air pressure for granted. The activities on the next few pages will demonstrate just how important air pressure is!

2.1 *Try These Yourself! Experiments with Air Pressure*

You Will Need

1. 1 4L plastic bag
   1 4L wide-mouth jar
2. 1 drinking glass
   1 piece of stiff card, large enough to cover the mouth of the glass
3. 1 bottle containing a drink (e.g. water or juice)
   1 one-hole stopper
   1 straw
4. 2 identical bottles
   1 one-hole stopper
   1 two-hole stopper
   2 identical funnels
   1 beaker or cup (for pouring)
5. 1 hard-boiled egg
   1 one-litre milk bottle
   paper matches
6. 1 test tube
   1 pan of water
   food colouring

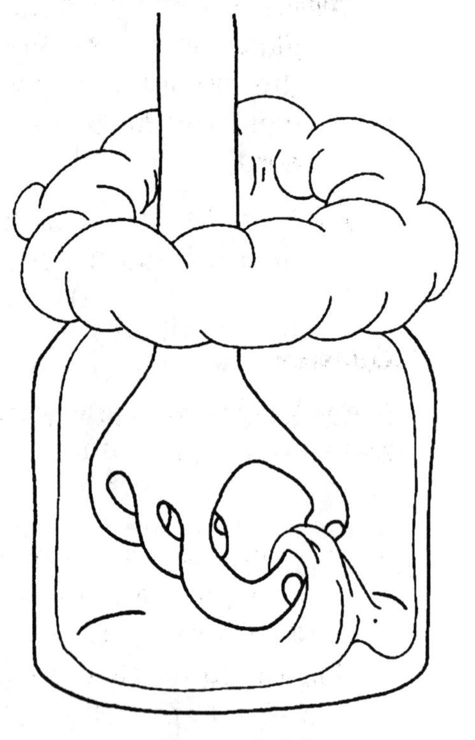

Figure 8 (a)

What to Do

1. (a) Put a 4-litre plastic baggie inside a wide-mouth jar of similar size, as in **Figure 8(a)**. Fold the sides of the bag around the outer edge of the jar, and tape it so that no air can leak in or out between the bag and the wall of the jar.

   (b) Try to pull the bag out of the jar, without tearing the bag.

Question

What happens when you try to pull the bag out of the jar? Why do you think this happens?

2. **(a)** Fill a drinking glass to the brim with water.

(b) Place a dry filecard over the top of the glass. Hold the filecard firmly against the rim of the glass, as you turn the glass upside down *over a sink* **(Figure 8(b))**.

(c) Now, remove your hand from the cardboard. What happens?

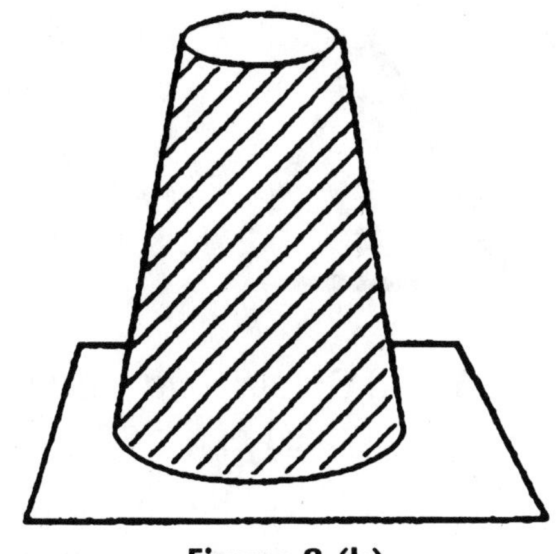

Figure 8 (b)

Question

What keeps the water in the glass when you take away your hand?

3. It is easy to drink water from a bottle, through a straw, the normal way. But try putting a clean, unused straw through a one-hole stopper, as in **Figure 8(c)**, then push the stopper into the mouth of the bottle. (A freshly washed plastic tube can be used instead of a straw.)

Now, try drinking through the straw.

Question

Why is it so difficult to drink through the straw when the straw is stuck through a stopper?

4. **Figure 8(d)** illustrates two identical bottles into which water can be poured through identical funnels. The bottle on the left has a two-hole stopper, while the one on the right has a one-hole stopper. Try pouring water into each of the bottles.

Question

In which case is it easier and faster to fill the bottle? Explain why this happens.

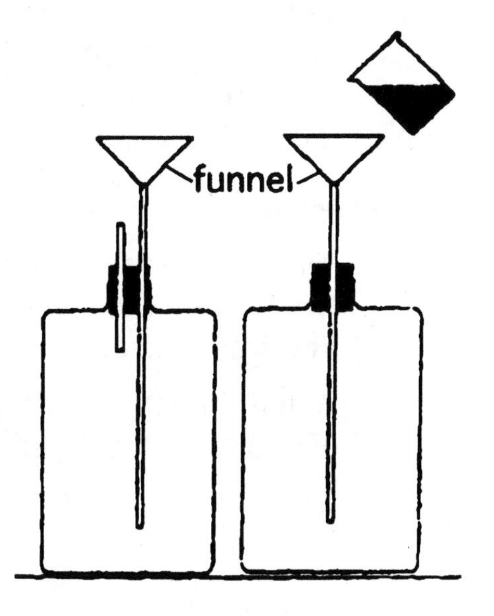

Figure 8(d)

CAUTION! A TEACHER MUST DEMONSTRATE THIS ACTIVITY. KEEP A CUP OF WATER HANDY, SO THAT A BURNING MATCH CAN BE EXTINGUISHED BY THROWING IT IN THE WATER.

5. (a) Obtain one peeled, hard-boiled, medium-sized egg.

   (b) Place the egg in the mouth of a 1-litre milk bottle, as in **Figure 8(e)**. You can see that the egg will not fall through the mouth of the bottle.

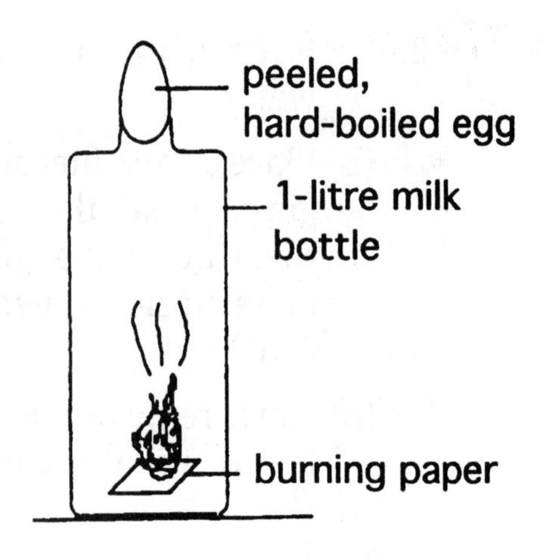

Figure 8(e)

   (c) Remove the egg temporarily. Roll up a small piece of dry paper, about 10 cm by 10 cm. Use a match to light the paper, and push the burning paper down into the milk bottle.

   (d) *Quickly* put the egg on top of the bottle.

   (e) Observe closely! At first, the egg will 'dance' at the mouth of the bottle. This happens because air warmed by the burning paper expands and forces it way past the egg and out of the bottle. Once burning stops, the air inside the bottle starts to cool off. Now, what does the egg do? Why do you think this happens?

   To remove the egg, turn the bottle over so that the egg sits inside the mouth of the upturned bottle. Run warm water from a tap over the bottle. After a short while, the warmed air in the bottle will expand and push the egg out.

6. Fill a large test tube completely with water, by putting it in a pan of coloured water. Keep the open end of the tube under water while you raise the bottom end of the tube out of the water, as in **Figure 8(f)**.

Figure 8(f)

Questions

1. Does the water stay in the tube when you lift its bottom end out of the water? Explain what happens.
2. What do you think would happen if you tried this experiment using a longer tube?

Think about It!

Think about what you have seen in these activities with atmospheric pressure when you answer these questions.

1. **(a)** Does atmospheric pressure just act downward, or does it act upward and sideward as well?

   (b) What evidence have you seen to support your answer?

2. When you open a can of tomato juice using a can piercer, why is it a good idea to punch two holes in the can rather than just one?

3. Why is the gas cap for a car not made perfectly airtight? What would happen if it was?

Figure 9

4. The boy in **Figure 9** filled a plastic pop bottle with water, then capped the bottle. He turned the bottle over and put the cap under the water in the bucket. He is now going to unscrew the cap. Predict what will happen to the water in the pop bottle when he removes the cap from the bottle. Explain what will happen. Why not try this experiment yourself?

Measuring Air Pressure

Daily weather forecasts always mention **atmospheric pressure**. Air pressure varies from day to day, sometimes from hour to hour. When the air is **humid** (moister than usual), a particular volume of air is actually lighter than when the air is dry.

Water vapour is made up of molecules of water (H_2O). If a particular volume of air contains a lot of water vapour, the water molecules take the place of some of the nitrogen molecules (N_2) and oxygen molecules (O_2) in the air.

Water molecules (H_2O) are actually ***lighter*** than nitrogen or oxygen molecules that make up most of the air. Therefore, if the air is humid, a particular volume of that air will be lighter than an equal volume of dry air.

Since moist air is lighter than dry air, the atmospheric pressure due to moist air will be less than the pressure due to dry air. When the weather forecaster says that atmospheric pressure is lower than usual, he will quite likely also say that the weather will be cloudy, possibly with rain or snow, because the air is more humid than on a 'high pressure' (dry air) day.

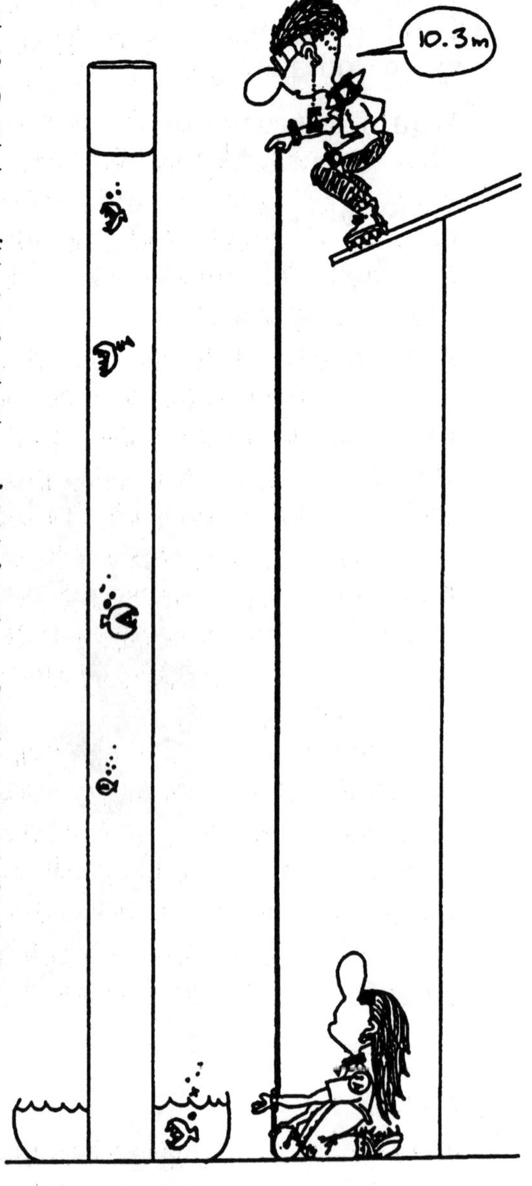

Figure 10

Remember what happened when you filled a test tube of water, then inverted it in a dish of water? (See Figure 8(f).) The water stayed in the upside-down test tube! It was kept up in the tube by air pressure. Air pressure on the water surface supplies enough force to balance the weight of the water in the tube, and this keeps the water up in the tube.

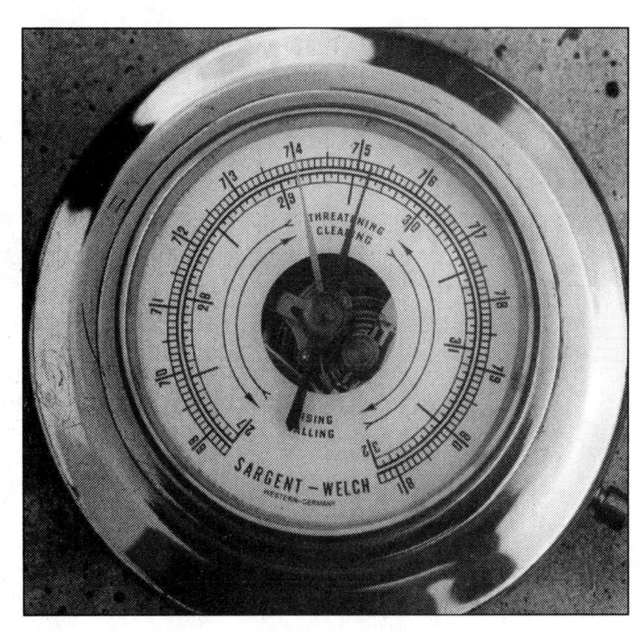

Figure 11

Atmospheric pressure will actually support a column of water 10.3 m high! If you had a very long tube, sealed at one end, you could fill it with water, then invert it into a large dish of water. **Figure 10** shows such a tube. Air pressure will support a column of water about as high as five basketball players standing on top of each other!

A tall tube of water, closed at one end, could be used as a barometer to measure atmospheric pressure. Obviously, it is not very practical to have a barometer 10.3 m high in your room. Scientists sometimes use barometers containing mercury instead. Mercury is a liquid metal, which is much more dense than water. On a normal day, atmospheric pressure will support a column of mercury approximately 76 cm high.

Mercury barometers are not practical for home or school use, because of the possibility of leakage or breakage. Mercury is a toxic substance and it must be handled with care. Home barometers are usually aneroid barometers. See **Figure 11**.

Aneroid barometers consist of an accordian-like metal container, which is partially 'crushed' by air pressure. A needle arrangement attached to the accordian-like metal box moves in response to changes in pressure acting on the container. Aneroid barometers are much smaller than mercury barometers.

Research

Find out how an aneroid barometer responds to changes in air pressure. If possible, examine one that has been taken apart.

SECTION 3 • Using Air Pressure

Your Lungs

Figure 12 illustrates one model of how air is pumped into your lungs. The two balloons represent your lungs, and the rubber sheet at the bottom represents your **diaphragm**. When you breathe in, muscles contract and cause your diaphragm to be pulled down. This creates a low pressure around your lungs, therefore atmospheric pressure outside your body forces fresh air (containing 20% oxygen) down into your lungs. When your diaphragm relaxes (moves upward), it raises pressure around your lungs, and forces 'used' air out of your body.

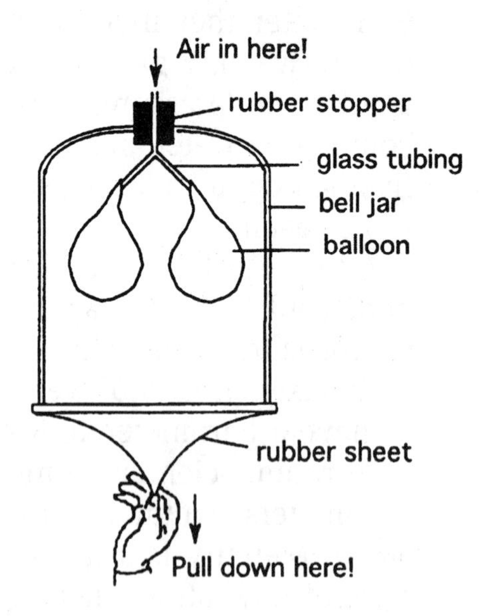

Figure 12

Project Idea: Make your own model, to show how a human lung works. It might be simpler to use just one balloon instead of two.

3.1 Try This Yourself! The Test Tube Submarine

What You Need

1 empty 2-litre plastic pop bottle

1 test tube

What to Do

Fill a 2-litre pop bottle to the brim with water. Fill a test tube with water and invert it just below the surface of the water in the pop bottle. Carefully pour out water from the test tube, until it is just about half full of water, then let it float in the water in the pop bottle. See **Figure 13**. If it sinks, start over again and use less water in the test tube.

When you have the half-filled test tube floating at the top of the bottle, cap the bottle tightly. Squeeze the bottle gently, and watch what happens to the test tube.

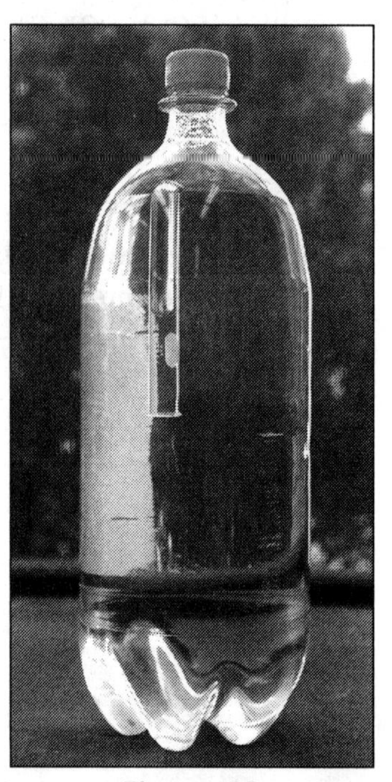

Figure 13

Questions

1. When you squeeze the pop bottle, what happens to the volume of the pocket of air trapped inside the test tube? Why does this happen?
2. Is the 'submarine' heavier or lighter when the pop bottle is being squeezed? Why?
3. What happens to the 'submarine' when you stop squeezing the bottle? What happens to the volume of the pocket of air inside the 'submarine'? Why does this happen?
4. Is the 'submarine' heavier or lighter when you stop squeezing the pop bottle? How can you tell?
5. When the volume of the air pocket inside the test tube becomes smaller, it is because something is compressing the air in the test tube. Where did this pressure come from?

Pascal's Law

Air and water are both fluids. (Gases and liquids can *flow*, so they are called *fluids*.) Over 300 years ago, French scientist Blaise Pascal discovered that *if you have a fluid in a closed container, and you change the pressure in any part of the container, the change in pressure will be transmitted everywhere throughout the container.* This fact of nature is called Pascal's Law.

The 'test tube submarine' in Figure 13 uses Pascal's Law. When you pushed on the walls of the closed container (pop bottle), the pressure you exerted was transmitted through the water (a fluid) and into the pocket of air (another fluid) inside the test tube. The increased pressure compressed the air inside the test tube into a smaller volume. The test tube now contained more water, so it became heavier and sank to the bottom of the pop bottle.

Pascal's Law is used in many modern devices. Some of the devices that use fluid pressure and Pascal's Law include automobile brakes, power steering, aircraft landing gear, the control surfaces on airplane wings, fuel pumping systems, earth moving machines, automobile lifts in garages, and many other applications.

3.2 *Try This Yourself! Using Fluid Pressure to Do Work*

You Will Need

50 cm of plastic or rubber tubing to join syringes

at least 2 plastic syringes, preferably an assortment of different sizes

What to Do

1. Connect 2 plastic syringes with about 1 m of plastic aquarium tubing. See **Figure 14**. Push in the piston of one syringe, and watch what happens to the other syringe. Try this with syringes of different size. Observe how far the pistons move in the two different syringes. Can you think of a way you might use this system?

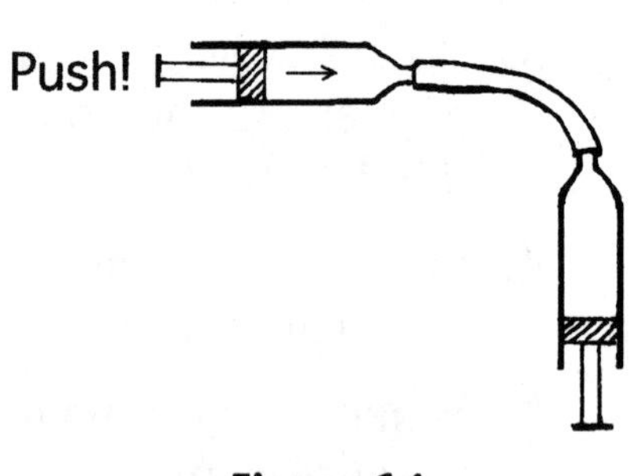

Figure 14

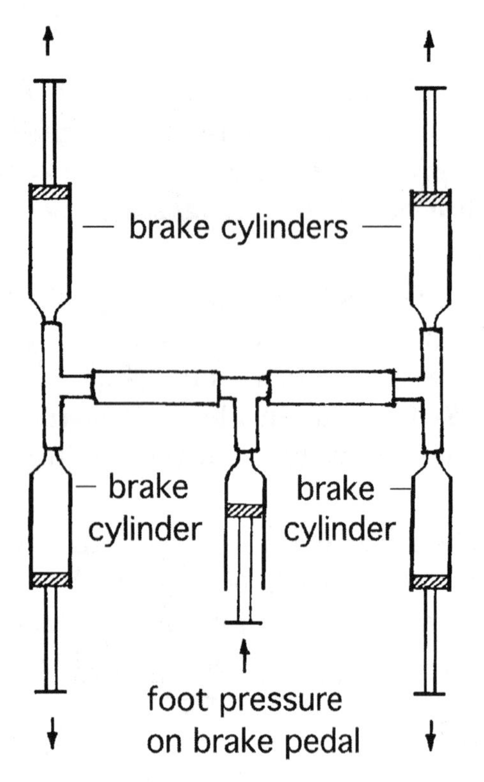

Figure 15

2. **(Optional)** Make a model of a car braking system, like the one in **Figure 15**.

   You will need 5 syringes and 3 T-junctions.

   Explain how your model works.

Using Air Pressure to Life Something Heavy

3.3 Try This Yourself! Can You Blow Up a Book?

What You Need

1 large freezer baggie
1 large book

Figure 16

What To Do

Place a heavy book on your desk. Now, try to blow the book off your desk. You can't do it this way! Now, place a clean 'baggie' under the book, as in **Figure 16**. Gather the end of the bag together and blow into the bag, the way you do when you 'pop' a grocery bag.

Questions

1. Why is it so easy to lift the book this way? How does Pascal's Law explain what you just did?
2. How might emergency rescue teams use Pascal's Law to lift a fallen tree off someone trapped under it?

A Class Project!

Arrange one table above a second table, as in **Figure 17**. Place as many clean, unused plastic sandwich bags ('baggies') as possible at equal spaces between the two table tops. Predict what will happen if all the students around the tables blow into these bags at the same time. Test your prediction!

Figure 17

Question

What happened to the table when everyone blew into the bags?

Extra Challenge! Have someone sit on the top table, and see if you can use your own air pressure to lift both the tables and the person!

SECTION 4 • Pressure in Moving Air

4.1 *Try These Yourself! Air Speed and Air Pressure*

What happens to the pressure in air, when the speed of the air is changed?

What You Need

1 piece of writing paper

2 empty plastic pop bottles

string

glue gun and glue sticks

stand with horizontal crosspiece

Figure 18

What to Do

1. Hold a piece of writing paper in front of your mouth, as in **Figure 18**. Blow hard over the top of the paper.

Questions

1. What happens to the paper?
2. Does the result you see suggest that air pressure is higher ***below*** the paper (where the air was not moving) or ***above*** the paper (where the air was moving faster)?

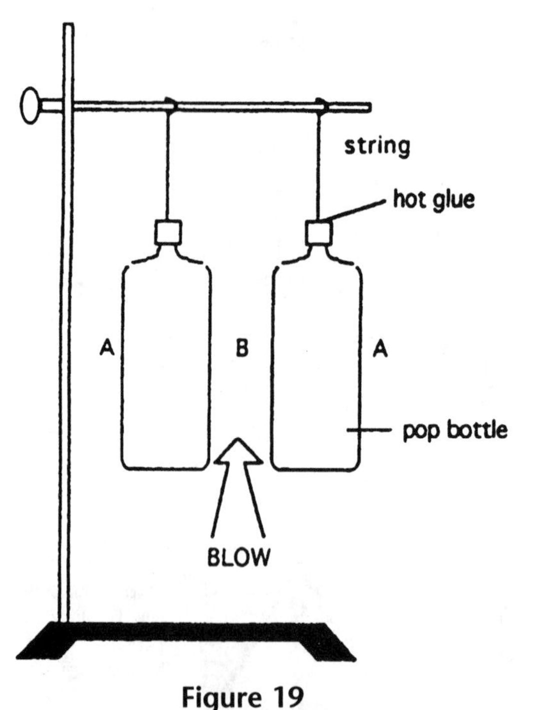

Figure 19

2. Hang two empty pop bottles by strings, so that they are about 5 cm apart, as in **Figure 19**.

(Use hot glue to attach the strings to the top of the caps.)

Predict what will happen if you blow hard between the two bottles.

Now test your prediction.

Question

Does the result suggest that air pressure is higher at **A** (where the air is still) or at **B** (where the air is moving fast)?

3. Place a ping pong ball in a freshly washed plastic funnel, as in **Figure 20**. Predict what will happen if you blow into the funnel.

   Can you blow the ping pong ball out of the funnel this way?

Figure 20

Challenge!

Can you make the ball stay in the funnel with the funnel upside down?

Question

Try to explain what happens.

Bernoulli's Principle

The experiments you have just tried illustrate an important principle discovered nearly 300 years ago by a scientist named Daniel Bernoulli (1700-1782). Bernoulli's Principle says that *where the speed of a fluid is high, the pressure is low; where the speed of the fluid is low, the pressure is high.*

Bernoulli's Principle explains many everyday occurrences. Here are a few examples:

Being Attacked by a Shower Curtain

Have you ever been attacked by your shower curtain? When the shower is running, the fast-moving stream of water creates a low pressure area between the shower curtain and you. The air outside the shower is at normal atmospheric pressure, and it pushes the curtain into the lower pressure area near you.

Passing Vehicles

When a large transport truck passes you on the highway, it sometimes feels like your car is being pulled toward the truck. Air is rushing between the two vehicles, creating a low pressure zone between your car and the truck. Higher pressure air on the other side of your car tends to push you toward the truck! A similar thing happens if two large ships pass too close to each other. Fast-moving water between the ships is at lower pressure than the water on the other sides of the ships. The ships are pushed together and may collide if they get too close to each other.

Perfume Sprayers and Garden Sprayers

Figure 21 illustrates how the Bernoulli Effect can be used to spray perfume from a bottle. When the bulb is squeezed, fast-moving air moves over the top of the tube that goes down into the perfume. This creates a low pressure at the top of the tube. Normal atmospheric pressure on the surface of the perfume pushes the perfume up the tube into the low pressure zone. The perfume is then carried out of the sprayer with the fast-moving air. Similar devices are used to spray paint and garden chemicals.

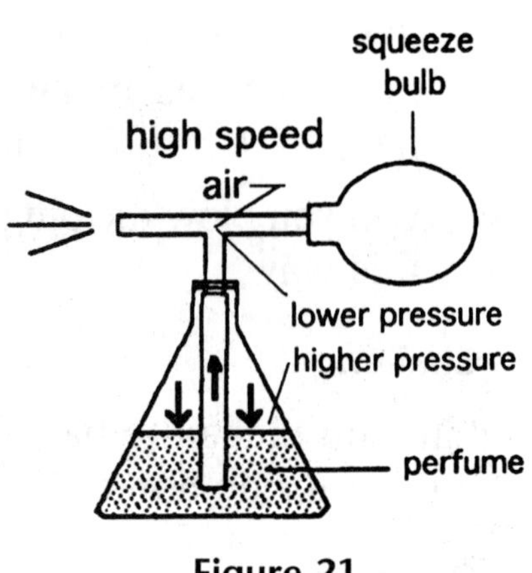

Figure 21

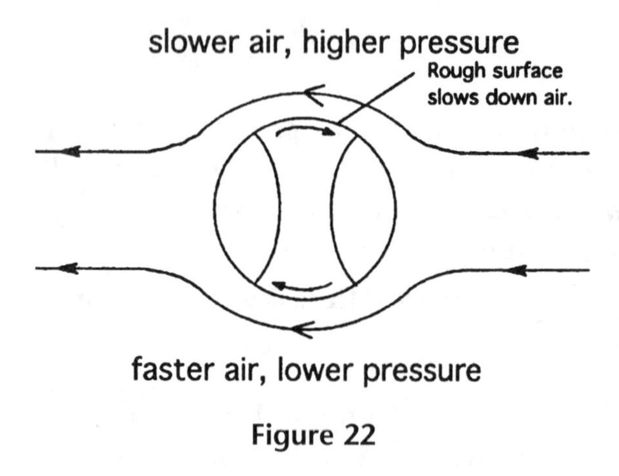

Figure 22

'Curve' Balls in Baseball

In **Figure 22**, a baseball is moving from left to right. It has been thrown with a spin. *You are looking down at the baseball.*

Imagine the baseball is stationary, and the air is moving past it from right to left. (This makes the explanation simpler.) At the top of the diagram, air is moving to the left and the ball is spinning to the right. The air moving past the ball is *slowed down* by friction. This slowing down of the air creates a zone of higher pressure. At the bottom of the ball on the diagram, air is moving to the left, but so is the ball. The speed of the air near the bottom of the ball is *higher* than the speed of the air near the top of the ball. Therefore, *air pressure is lower at the bottom* of the ball on the diagram. The difference in air pressures results in the ball being pushed down on the diagram. (From the point of view of the pitcher, the ball will curve to the right.)

Bernoulli's Principle explains other effects seen in sports. A golf ball will 'hook' (curl to the left, for a righthanded golfer) or 'slice' (curl to the right for a righthanded golfer) depending on the kind of spin imparted to the ball when it is struck. In table tennis, overspin imparted to the ball will make the ball 'dive' more quickly over the net. In soccer, a ball kicked with a sidespin will also 'hook' or 'slice,' depending on the spin put on it.

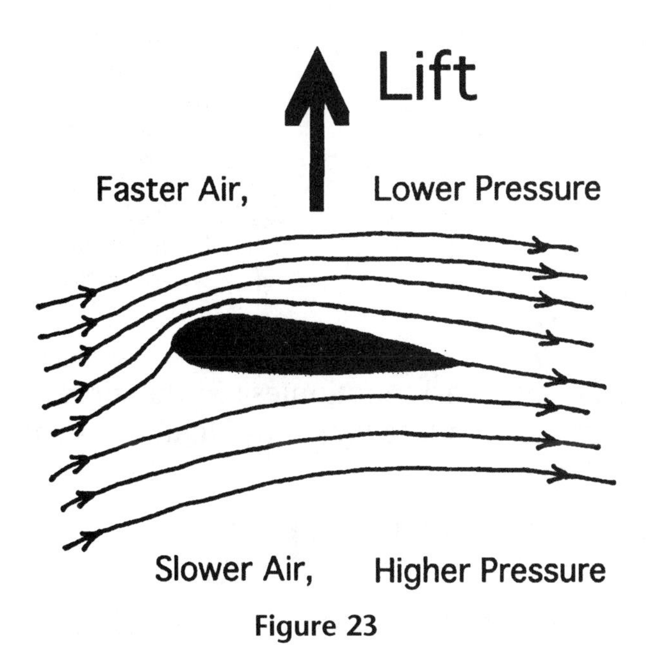

Figure 23

Air Foils

The main reason there is lift on an airplane wing is its typical airfoil shape, seen in **Figure 23**. This shape, and the angle the wing makes with the direction of travel, causes air moving above the airfoil to move faster than air moving under the airfoil.

The difference in air speeds over and under the wing causes a difference in pressure, and this creates an upward force called **lift**. If the lift force equals or is greater than the force of gravity on the plane, **flight** becomes possible.

Figure 24

Don't try this at home! In **Figure 24** a 'wing walker' holds on for dear life, as the pilot of this aircraft performs at an air show. A powerful single engine provides forward thrust, while the airfoil shape of the wings provides **lift** and keeps **drag** to a minimum. (The unusual 'passenger' does not help to keep drag as low as possible.) The **force of gravity** pulls downward on the aircraft and passenger.

SECTION 5 • Flight

Go Fly a Kite!

Forces on a Kite in Flight

If a suitable wind is blowing, a kite will 'fly' nearly motionless in the air, even though it is heavier than air. Three forces act on a kite that is in flight. These are shown in **Figure 25**.

1. The wind, deflecting off the bottom surface of the kite, exerts a force on the kite called the **aerodynamic force**.

   The aerodynamic force not only provides an upward **lift** to the kite, but also causes drag. (The drag force acts in a direction opposite to where the kite is pointed.)

2. The **force of gravity** pulls down on the kite.
3. The string exerts a force on the kite. When the **force exerted by the string**, the **aerodynamic force**, and the **force of gravity** balance each other, the kite hovers in the same place in the air.

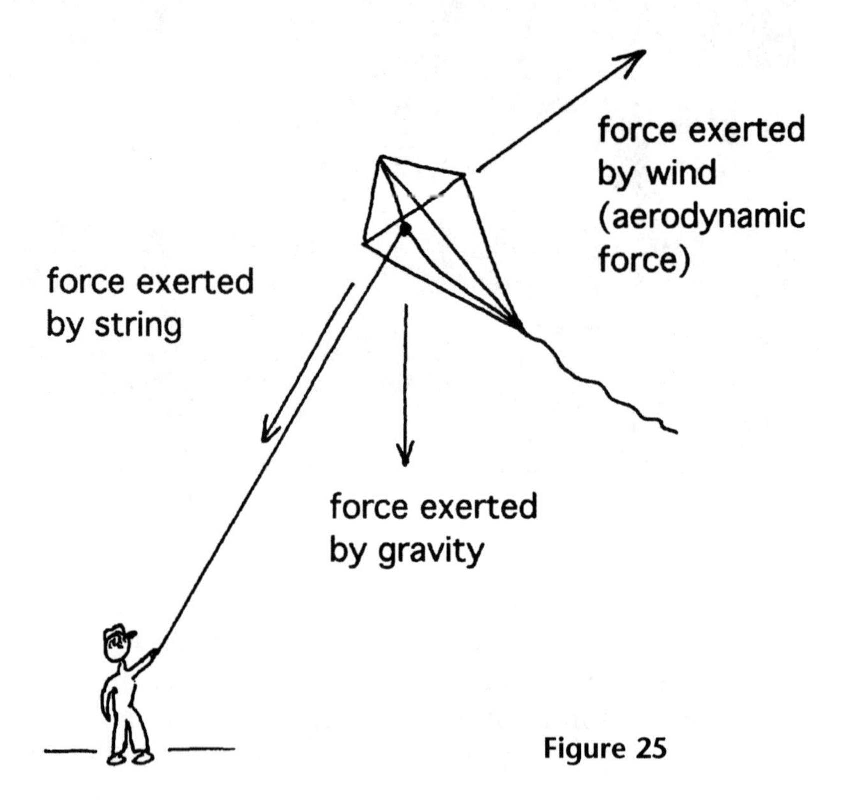

Figure 25

In **Figure 26** a hang glider gains forward speed using gravity by falling from a high starting point. Instead of relying on the wind for lift, as a kite does, the glider moves through the air, and experiences the same lifting effect as a stationary kite in the wind.

Figure 26

Forces on an Airplane in Flight

The main source of **lift** on an airplane wing is the difference in air speed over and under the wings. (Remember **Bernoulli!**) Also, wings are tilted slightly upward, and air is actually pushed *downward* by the wing surface. The deflected air exerts an equal and opposite force, called the **aerodynamic force**, *upward* on the wing. The **aerodynamic force** does two things to the wing:

1. It helps provide **lift**; and
2. it creates **drag**, which acts in a direction opposite to the direction in which the airplane is moving. Drag tends to slow down an airplane.

Figure 27 illustrates the forces exerted on an airplane in flight.

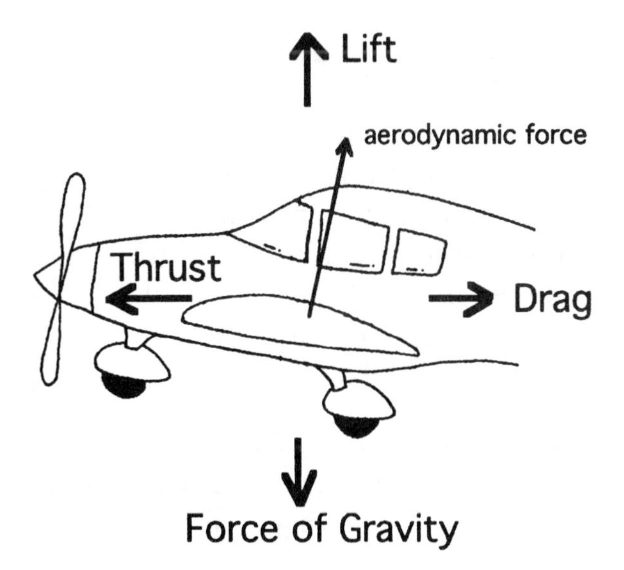

Figure 27

Figure 28(a)

Drag is made as small as possible, and lift as large as possible, by selecting airfoil shapes that work best for the range of speeds at which a particular airplane will travel. For an unpowered glider, the shape that works best is shown in **Figures 28(a)** and **(b)**. The wings are very long and quite narrow. This combination works best for a slow-moving glider, where the only thrust force is gravity.

High-speed jet aircraft require different types of wing shapes and sizes. The forward **thrust force** may be supplied by propellers, jet engines, or rocket engines. Of course, the greater the thrust force is, the more drag there is. Drag must be reduced by choosing the most appropriate shape for the wings and body of the aircraft.

Project Idea

Make (or examine) a collection of photographs of various types of passenger and military aircraft. Look for features that are different in the design of the aircraft, especially their wings. Compare the shapes of the airfoils as seen from the side, from above and from below. What type of aircraft have wings that are swept back? What do the wings look like on a very high speed (supersonic) military jet?

Figure 28(b)

Controlling the Motion of an Airplane

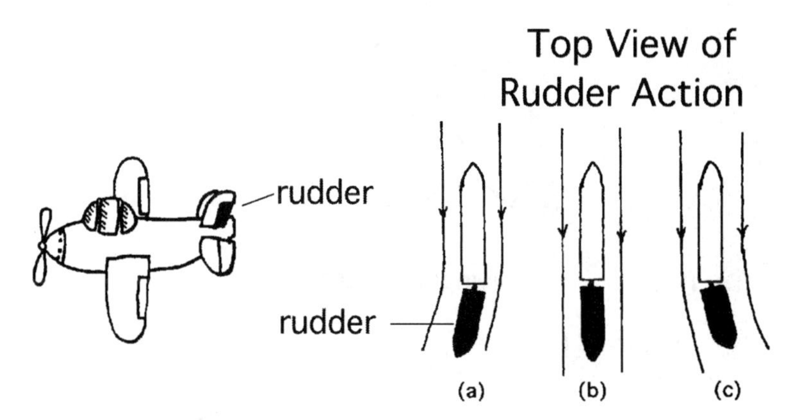

Figure 29

Turning Right or Left

An airplane can be made to turn left or right in two ways. **Figure 29** illustrates how the rudder of an airplane moves. It works in a way similar to the rudder of a boat. In **Figure 29(a)**, air deflecting off the rudder will push the tail end of the plane to the right, and cause the nose of the plane to go to the left. In **Figure 29(c)**, air deflecting off the rudder pushes the tail end of the plane to the left, and causes the nose of the aircraft to turn to the right.

Another way to make a plane turn left or right is use the plane's **ailerons** (pronounced ale-er-on) to make one wing lift up while the other wing dips down. As **Figure 30** shows, the ailerons are hinged flaps located near the ends of the wing.

To make the plane **bank** to the left, the left aileron is hinged upward, which will make the left wing 'dip'. The right aileron is hinged downward, which makes the right wing lift up. The effect of the ailerons in this situation is to make the airplane bank to the pilot's left. The pilot also uses the rudder to help complete the turn. To bank to the right, the left aileron is hinged downward, and the right aileron upward.

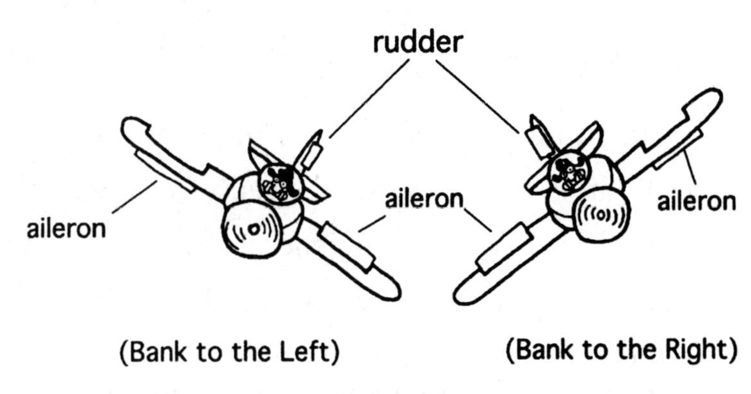

Figure 30

Going Up or Down

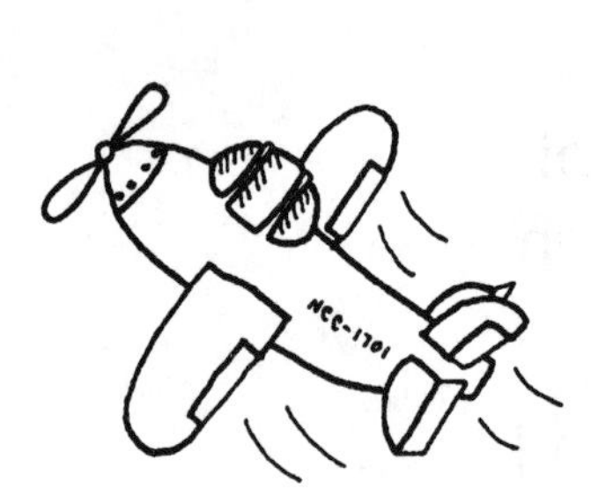

Plane Climbing: Elevator Up   Plane Diving: Elevator Down

Figure 31

At the tail end of an aircraft, there are horizontal fins, called **stabilizers**. See **Figure 31**. On the stabilizers, there are hinged flaps called **elevators**. If the pilot makes the elevators hinge upward, this will cause the tail end of the airplane to drop, and the nose to lift. If the pilot makes the elevators hinge downward, this will cause the tail to lift, and the nose to drop.

If the speed of the aircraft increases, the air rushing past the wings automatically provides greater lift, and this will tend to make the aircraft increase its altitude. If the airplane slows down, lift decreases, and this causes the plane to lose altitude.

A pilot has to take many factors into account when keeping his airplane in flight and controlling its direction and altitude.

Research

Investigate one or more of the following topics in detail. Make a wall chart to show your findings to the rest of your class.

1. How do the wing shapes of airplanes that travel faster than the speed of sound (supersonic) differ from the wing shapes of small, slower aircraft?
2. If you look out the window when a commercial passenger airplane is landing, you see a lot of interesting things happening on the wings. Find out what these parts of the airplane are, and what they do:
   air brakes, spoilers, wing flaps, and trim tabs.
3. What is meant by 'stall speed'? What causes a plane to 'stall'?
4. In what ways is a propeller similar to a wing?
5. How does a helicopter work? How is it steered?
6. Read about the Hawker-Siddeley Harrier 'jump-jet' aircraft. What special features does it have that other aircraft do not have?

Paper Airplane Contest

5.1 Try This Yourself! A Class Project

What You Need

1 piece of plain copy paper
1 pair of scissors
1 tape measure
1 stop watch

What To Do

Have a contest to see which student or which team of students can design and build a paper airplane that will

(a) travel the farthest distance in a straight line, or

(b) stay in flight for the longest time.

(All aircraft must be launched from the same starting point, by the same person.)

Suggestion

Before you start this contest, you might do some research in the library or in other sources, to obtain ideas for paper airplane design. Your class might like to set some common rules about size of paper, use of glue or tape, and so on.

First, have some trial runs, to get a good idea of which designs work the best. When each individual or team is ready, have a serious contest in one or both of the categories (a) and (b).

Keeping Track of Your Airplane's Performance

(a) Table 1: Distance Travelled in a Straight Line

Trial Number	Distance Travelled (metres)
1	
2	
3	
4	
5	
6	
7	
8	
9	
10	
Average	

Launch your paper airplane from a chosen spot and measure how far it travels, in a straight line, in metres. Repeat this nine more times, and each time record the distance travelled. Record your measurements in a copy of **Table 1**.

Calculate the average distance your plane travelled for the ten trials. To do this, add up the ten measurements, then divide the total by ten.

Prepare a chart like **Figure 32**, to show your results graphically.

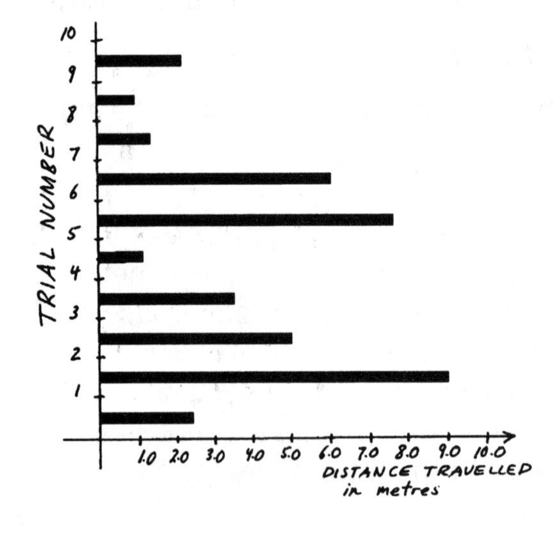

Figure 32

Who Wins the Contest?

Your group can make the rules. You can decide whether the winner is the airplane that averages the greatest distance over ten trials, or the one that travels the greatest distance in its best single trial. Or, you might decide to use the average of the best three, or best five trials.

(b) Table 2: Time in Flight

Trial Number	Time of Flight
1	
2	
3	
4	
5	
6	
7	
8	
9	
10	
Average	

In **Part (b)**, the purpose is to design a paper airplane that will stay in flight for the longest time. Once more, have some trial runs, and when everyone is ready, have a contest to see who can make the paper airplane that glides for the longest time. Use a stop watch to time the flight of your airplane. If time permits, do ten trials, and record your measurements in a copy of **Table 2**. Also, design a chart to display your results. It will look like **Figure 32**, but you must choose a suitable scale to show the range of times you measured.

Questions

1. Of the paper airplanes that travelled the greatest distance, what common features did you observe in their designs and shapes? Describe or sketch the most successful airplane.
2. Of the airplanes that stayed in flight for the longest times, what common features did you observe in their designs and shapes? Describe or sketch the most successful airplane.

SECTION 6 • Other Ways to Fly!

Hot-Air Balloons

Figure 33 (a) (b)

A hot-air balloon **(Figures 33** and **34)** works differently than an airplane. To lift off, the balloon and its passengers must literally 'float' in the air, just as a boat floats on water. A boat is buoyed up by pressure exerted by the water around and below it. A hot-air balloon is buoyed up by pressure exerted by the air around and below it.

The hot-air balloon structure is, by itself, heavier than air. However, if air inside the balloon is warmed up using a torch, as in **Figure 33(b)**, the air expands and becomes less dense than the cooler air outside the balloon. The hot-air balloon will now 'float' on the cool air around it.

6.1 Try This Yourself! A Garbage Bag Balloon

What You Need

1 large black garbage bag

string or a twist tie

What to Do

To see what happens to the volume of a certain mass of air when it is warmed, partially fill a black plastic garbage bag with air by dragging it through the air, then tying the bag closed. Warm the air inside the bag either by setting it out in the sun, or by blowing warm air from a hair drier on it. Observe what happens to the volume of the bag.

Questions

1. Has the volume of the air inside the balloon changed?
2. Has the mass of the air inside the balloon changed?
3. Density is a measure of how much mass there is in a certain volume of air. Is the air in the warmed garbage bag more dense than before, less dense than before, or the same density as before? Explain why.
4. A student tried blowing a soap bubble outside on a cold day. She noticed that the bubbles tended to rise quickly through the cold air and float away. Why did this happen?

Figure 34 Hot Air Balloons in Grande Prairie, Alberta

Dirigibles

Another way to achieve flight with a balloon is to use a sealed balloon that is filled with a gas that is less dense than the air in the atmosphere. **Helium** is much less dense than air, and it is safe to use because it does not burn. Dirigibles (also called **blimps**), which you sometimes see at large sports events, are filled with helium. Early attempts at lighter-than-air flight used **hydrogen** because of its low density. Unfortunately, hydrogen is extremely flammable and unsuited for balloons that are designed to carry passengers. A single spark can cause the entire balloon to explode.

Research

Read about the disaster of the Hindenburg, a huge, passenger-carrying German dirigible built in 1936 that was filled with hydrogen gas.

Parachutes Are a Real Drag!

Figure 35

A sky diver is quite pleased to have plenty of **drag** on his parachute. Without it, he would not slow down enough to survive the landing **(see Figure 35)**!

A parachute looks a lot like an umbrella. Its purpose is to slow down the descent of a person or thing through the atmosphere. Parachutes can be used to allow a pilot and crew to escape safely from a disabled airplane (or hot-air balloon). They are also used by the military to drop supplies and people to a battlefield in wartime. Parachutes have been used to slow down space capsules returning to earth from orbit, or to slow down fast-moving airplanes that are landing on runways that are too short for a normal landing.

The parachutist in the photograph **(Figure 35)** is part of a Canadian Forces demonstration team called the Sky Hawks. When a sky diver first jumps out of a plane, gravity pulls him or her down, but the drag force exerted by the air, which opposes the force of gravity, increases with speed. The drag force allows the sky diver's falling body to reach a maximum or 'terminal' speed of about 190 km/h. But when the nylon parachute opens, the parachutist experiences a sudden 'jolt', and his or her body is rapidly slowed to a mere 15 to 25 km/h.

The rate of fall will depend on the force of gravity on the parachutist, the design and diameter of the chute, and on the altitude (since air is more dense at lower altitudes). Landing speed in a normal situation is about the same as it would be if someone jumped off a ladder 1.5 m high. Don't try this at home!

6.2 *Try This Yourself! Make Your Own Parachute*

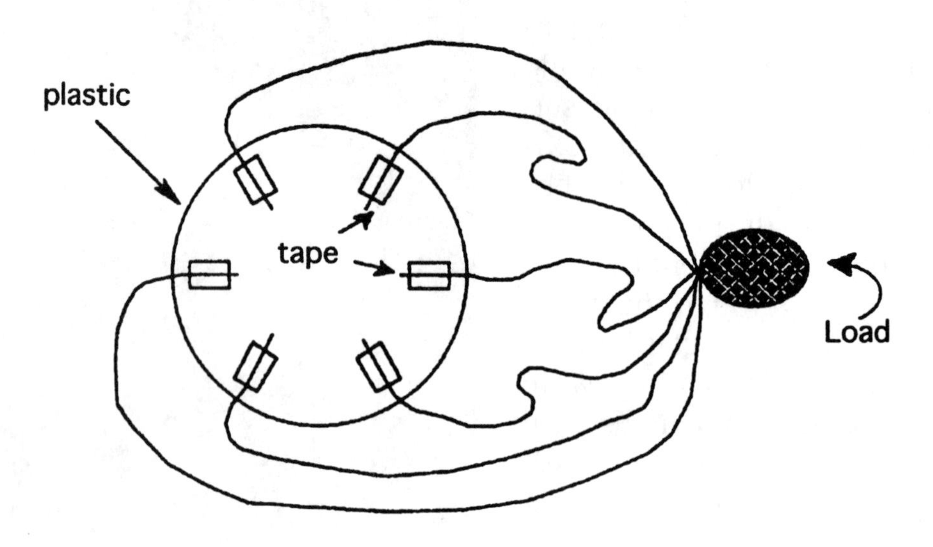

Figure 36

What You Need

2 plastic garbage bags

metre stick

string

scissors

masking tape

a bundle of washers for a 'load'

or a toy 'parachutist' (e.g. a small teddy bear)

a safe place from which to drop a parachute

What to Do

1. Prepare a model parachute like the one in **Figure 36**. Cut a circle out of garbage bag material, about 30 cm diameter.
2. Attach about six 40-cm strings, at equal distances, around the outside of the circle. Use masking tape.
3. Attach the other ends of the strings to the model parachutist (or a bundle of washers).
4. Try dropping the parachute and its 'load' from a safe height and, if possible, measure the time it takes to hit the ground. (You might try dropping an identical 'load' at the same time, but without a parachute.)

Question

When a sky diver opens his or her parachute, does the parachute slow down gradually or suddenly? Explain why this happens.

Challenges!

Try making the parachute surface with a circle that has twice the radius. How does this affect the rate at which the parachute and its load fall? Try one with three times the radius.

Have a contest to see who can design a parachute that will drop a particular load at the slowest rate from a given height.

Could You Fly on the Moon?

You may have seen pictures of astronauts walking on the moon. They have to wear spacesuits because there is no air on the moon. An astronaut walking on the moon is surrounded by a vacuum!

There is no air on the moon because the force of gravity exerted by the moon is too weak to hold an atmosphere. If there were molecules of air or water vapour on the moon's surface, they would escape the moon's weak gravity and go off into space.

The force of gravity on an astronaut on the moon is about one-sixth of what it is here on earth. An astronaut feels much, much lighter on the moon, and even ordinary walking requires some practice in this strange environment.

If there is no air around the moon, airplanes and kites will not work, because there is no air to supply lift to their wings. Balloons will not rise, because there is no air in which they can float. *A balloon on the moon's surface is similar to a boat on earth at the bottom of a dried-up lake!*

One way to obtain 'lift' on the moon is to use rocket engines. Rocket engines actually work better in a vacuum, because exhaust gases coming from the rocket are slowed down if they have to pass through air outside the rocket. Very little thrust is needed from the rocket engines to escape the moon's weak gravitational force.

Questions

1. Would a parachute work on the moon? Explain your answer.
2. When astronauts left the moon to return to earth, the space vehicle and the rockets they used were much smaller than the vehicle and rockets they needed to leave the earth and go to the moon. Why is this possible?

Research

1. Find out if there are any animals that use their bodies as parachutes, to 'fly' over short distances. Prepare a wall chart that describes what you learned.
2. Many seeds are carried by the wind. Find out how dandelion seeds are spread over your community.
3. What gas is used in weather balloons? What sort of information is gathered using weather balloons?
4. Find out what Leonardo da Vinci predicted a parachute might look like, long before humans had learned how to fly by any means.
5. Investigate how various kinds of birds — eagles, sparrows, owls, geese, or hummingbirds — have different wing shapes for different types of flight, and different purposes. Make a wall chart showing what you have learned.
6. How do insects of various kinds fly? How rapidly do their wings beat? How are they able to make their wings beat so rapidly?
7. Investigate a topic related to this book that is of particular interest to you, personally. Write a brief report, or prepare a wall chart to share what you learned with your classmates.

SECTION 7 • Test Your Knowledge!

Use the list of Key Words to complete a copy of the crossword below.

Please do not write in this book.

Key Words

area	**elevator**	**lift**	**oxygen**
air	**force**	**low**	**pascal**
barometer	**gravity**	**lung**	**pressure**
Bernoulli	**high**	**mass**	**thrust**
dense	**inertia**	**nitrogen**	**volume**

Cloudy weather often results when atmospheric pressure is (**1 across**).
Atmospheric (**7 across**) forces air into this part of your body (**1 down**).
This control surface on an airplane is used for changing altitude. (**2 across**)
If something is matter, it occupies space and we can measure its (**3 down**).
Pressure is a measure of how much (**4 down**) acts over a certain (**13 across**).
This scientist's principle explains how an airfoil achieves lift. (**5 across**)
Some balloons achieve lift because they are filled with hot (**6 across**).
Anything that is matter has (**8 across**). This also means it has (**11 across**).

If a lot of force acts on a small area, the pressure will be **(9 down)**.

Pressure is measured using a device called a **(5 down)**.

Pressure is measured in a unit called the **(7 down)**.

The downward force that acts on an airplane is the force of **(10 down)**.

Drag slows an aircraft, but the forward **(16 across)** from the engines and the upward **(14 across)** from the airfoil help keep the airplane in flight.

Air is made up of about 80% **(12 down)** and 20% **(15 across)**.

A hot air balloon floats in air, because the warm air inside it is less **(17 across)** than the cooler air surrounding the balloon.

Show What You Know!

Here is an opportunity to show others what you know, and have some fun doing it. Your task is to prepare a demonstration that clearly shows one of the ideas you have learned in this book. Working individually or in small groups, design a demonstration to show one or more of the facts or ideas listed below. If you work as a group, you should prepare several demonstrations. When your demonstrations are ready, you can put on a science show for your classmates, for other classes, or for your family and friends.

1. Air has mass.
2. Air occupies space.
3. Air has weight.
4. Air exerts pressure in all directions — down, up, and sideways.
5. Air pressure is measured with a barometer.
6. Air pressure may change from hour to hour.
7. We use air pressure to breathe, to drink through a straw, to pump water from a well, to operate a sink plunger, and .... ?
8. If the pressure is changed at any point in an enclosed fluid, the change in pressure will be felt everywhere in the enclosed fluid. (Pascal's Law)
9. Pascal's Law can be used to lift heavy loads with very little effort.
10. Where the speed of a fluid is high, the pressure is low; where the speed of the fluid is low, the pressure is high. (Bernoulli's Principle)
11. Bernoulli's Principle explains many things: lift on an airplane wing, the curving motion of a spinning ball (golf, baseball, ping pong, soccer, basketball), perfume sprayers, etc.
12. Four forces act on an airplane in flight: lift, drag, thrust, and gravity.
13. Aircraft have control surfaces that allow the plane to change altitude or direction. They include elevators, ailerons, and rudder.

14. Hot air balloons and helium balloons 'float' in the denser air around them.

15. Parachutes must provide enough drag to slow the fall of a parachutist.

Challenge!

Figure 37

Find out how the tiny hummingbird in **Figure 37** achieves flight. How rapidly must its wings beat for it to remain hovering in its search for nectar from a flower (or a hummingbird feeder)?

Other Science and Technology titles from Trifolium Books

SPRINGBOARDS FOR TEACHING SERIES

NEW

INVENTEERING: A Problem-Solving Approach to Teaching Technology

Bob Corney & Norm Dale

An essential "getting started" resource for teachers of **Grades 1–8** wanting to provide their students with hands-on technological experiences.

8½" x 11" • 128 pages • Soft cover • Illustrations
ISBN: 1-55244-014-1 • $29.95 Can. • **Available**

IMAGINEERING: A "Yes, We Can!" Sourcebook for Early Technology Experiences

Bill Reynolds, Bob Corney, and Norm Dale

Packed with ideas to stimulate young students' imagination and creativity as they explore the issues and applications of technology. For teachers of **Grades K–3.**

8½" x 11" • 144 pages • Soft cover • Illustrations
ISBN: 1-895579-19-8 • $29.95 Can. • **Available**

ALL ABOARD!: Cross Curricular Design and Technology Strategies and Activities

By Metropolitan Toronto School Board teachers

This teacher-tested resource helps educators integrate design and technology easily and effectively into day-to-day lessons. For teachers of **Grades K–6**.

8½" x 11" • 176 pages • Soft cover • Illustrations
ISBN: 1-895579-86-4 • $21.95 Can. • **Available**

Take a Technowalk to Learn about Materials and Structures

Peter Williams & Saryl Jacobson

Provides teachers of **Grades K–8** with 10 fun Technowalks designed to encourage students to investigate the **materials and structures** that surround us.

8½" x 11" • 96 pages • Soft cover • Illustrations
ISBN: 1-895579-76-7 • $21.95 Can. • **Available**

NEW

Take a Technowalk to Learn about Mechanisms and Energy

Peter Williams & Saryl Jacobson

Now teachers of **Grades K–8** have fun with 10 new Technowalks designed to encourage students to investigate **mechanisms and energy** in the classroom, school and community.

8½" x 11" • 92 pages • Soft cover • Illustrations
ISBN: 1-55244-004-4 • $25.95 Can. • **Available**

TEACHERS HELPING TEACHERS SERIES

BY DESIGN
Technology Exploration and Integration

By the Metropolitan Toronto School Board Teachers

Over 40 open-ended activities for **Grades 6–9** integrate technology with other subject areas.

8½" x 11" • 176 pages • Soft cover • Illustrations
ISBN: 1-895579-78-3 • $39.95 Can. • **Available**

Mathematics, Science, & Technology Connections

By Peel Board of Education Teachers

Twenty-four exciting integrated Math, Science, and Technology activities for **Grades 6–9**.

8½" x 11" • 160 pages • Soft cover • Illustrations
ISBN: 1-895579-37-6 • $39.95 Can. • **Available**